500

FROM BOURGEOIS LAND

FROM BOURGEOIS LAND

Iain Crichton Smith

LONDON
VICTOR GOLLANCZ LTD
1969

575 00271 9

PRINTED IN GREAT BRITAIN
BY EBENEZER BAYLIS AND SON, LIMITED
THE TRINITY PRESS, WORCESTER, AND LONDON

ACKNOWLEDGEMENTS

Two of the poems in this collection have previously appeared in print. Acknowledgement is due to *The Times Literary Supplement*, which published No. 21 ('At the Sale'), and to *Transatlantic Review*, which published No. 37.

1

One might still argue that it's possible
to conceive a mind that sees no wrong in this
accountancy of murder, smells no gas,
does not remark how naked bodies topple
stripped of rings and watches into the abyss.

That some are gifted with imagination
and some are not, but have a real flair
for abstract work such as assiduous care
for numbers, for timetabling migration,
for calculating train-times and the fare.

That scrupulous conscience can assume the form
of being careful about death itself
—lest any should escape—and feel deep grief
if clerkly error should let Jews confirm
that God might really care for human life.

Perhaps it's possible to see him drive
from his bourgeois home to his office, bright with glass,
sit at his desk, consider how (alas)
the times are changing, it's imperative
to get some typists that are not so crass,

pencil an entry in his calendar,
a party or some fishing or some golf,
a spot of carpentry (say that kitchen shelf)
a game of chess, an evening at the fire
with his wife and/or his children and himself.

Or, rolling sleeves up, level out the lawn
on a day so clear that Poland can be seen
or even Russia, and the houses shine
row on row in sunlight and Herr Braun
worships, in a garage, his machine.

Perhaps it's possible to see him sign
page after page, a clerkly scrupulous man
who takes a pride in neatness, in a plan
which guides with accuracy a loaded train
across vast wastes as down an autobahn.

But this demands firmer imagination
than he himself expended on the dead
and seems so monstrous and so nearly mad
that it is simpler and more close to reason
to say at last when everything has been said:

"Such human beings are conceivable
if, being dead, one can be said to live,
if to the plague I answer "I forgive",
if to the stone "You are forgivable",
if one could say, "I will forgive you, grave."

But otherwise, you clerk who hated error
more than the sin that yet involves us all,
I say, "You are so monstrous I would call
the bells of hell, gassed faces in the mirror,
to enliven age on age your bourgeois soul."

2

Entering your house, I sniff again
the Free Church air, the pictures on the wall
of ministers in collars, all these dull
acres of brown paint, the chairs half seen
in dim sad corners by the sacred hall

under the spread antlers of that head
mildly gazing above leathern tomes.
So many draperies in so many rooms.
So many coverlets on each heavy bed.
A stagnant green perpetuates these glooms.

And then the stairs. The ancient lamps and the
scent of old prayers, texts of "God is Love"
Did any children grow through all this grief?
The ceilings seem to sigh, the floor to be
carpeted by a threadbare dim belief.

Such pressures on the head. And then I see
in an oval frame an eighteen-year-old girl
like Emily Bronte staring from the peril
of commandments breaking round her. And I pray
that she was happy, curl on winding curl,

even though I see the stains around her face,
the ancient tints of brown that eat at brow
and hair and nose, and make me see her now
as almost rusted in this world of grace.
How little beauty conscious sins allow!

I enter the great garden with its red
and dripping roses, laburnums and the tall
tulips and the columbines, the small
and holy Rose of Sharon. All the dead
are lost in colour as the dewdrops fall.

I watch a bee nuzzling from tower to tower
of brilliant yellow, each with soundless bell,
its hairy body busy in the smell
and light of evening. From flower to flower
it flies and sucks, quivers, then is still,

so gross and purposeful I can forget
the tall and simple flowers it feeds on here
in this bright garden of a freer air.
I pray that she, some gross and fruitful night,
under less heavy coverlets as bare

as these tall flowers, allowed new life to start
from her body's honey, turning to the wall
that portrait of her father, stern and tall.
And that the Rose of Sharon at her heart
quivered and quietened in her radiant fall.

3

SPEAKER

Ah, well, let the sun shine on you
as it does on a brass plate,
as it does on a frog's throat,
as it did on Demosthenes.

And as the local press
write down every word,
as the snap comes out unblurred,
perch on your glowing toes.

For local lead becomes gold.
To speak the truth is more than bad taste.
It is a treachery to the provost,
to the bronze-winged angel with the lead shield.

With every word you now orate
perpetuate the houses of our fathers
built of lies, deceit, and sunny weathers
that hide the sooty rafters of our hate.

Why should you not appear so fresh and clear?
New thought's disturbing and is not for us.
Better by far the tender avenues,
the shady trees, the undistracted air.

Why should you bring a nightmare to their dream
or trouble elders with unsettling clouds?
Why should you put ideas in old heads?
Or devil's dance in a harmonious limb?

It's not your task. Rather reflect such light
as shines around you and go down in peace
to an honourable age, a better press
than some received who spoke what's true and right.

4

Poets aren't dangerous, you think.
They have no fruitful honours in their keeping
nor any terrors that may mar your sleeping
nor medals, orders, offices or rank.

They walk on clouds. They sing to themselves alone.
Poetry is their hobby, nothing else.
They hear at evening metaphysical bells.
You, on the other hand, have a public phone.

They stutter awkwardly when asked to speak
and when they speak they lose their audience.
Masters of words, they have no resonance.
Where you are simple, often they're oblique.

But, let me tell you, poets can destroy.
Your suit is empty when our arrows fly.
The worst of dangers travel from the sky
as plagues and rockets, planes and birds of prey.

And, cruising here, drawing my life from clouds,
I watch such gestures as a clown might make
arrayed in time and skin like an old sack,
shapeless below the venom of such birds

as I am member of, impersonal, grave,
the Parcae of hard honour, single-beaked,
immune to bribery, however weak,
and trained to justice by implacable love.

5

DIDO AND AENEAS

What did the papers say of that romance
between Dido and Aeneas? How he left
his coloured bride and sailed on his suave craft
to find a country and to found a province.

That Trojan puritan, immune to love,
believing in his gods, adoring duty,
abandoned on a morning his black beauty
and saw the smoke like signals from her grave.

Accord him honour, he had work to do,
set up committees, build his marble halls,
train a Civil Service and install
a vulgar praetor of a parvenu.

Teach fountains to tell stories of a Rome
which would explore the world, subdue old races,
decorate by plagiarism vases,
and cultivate the virtues of the home.

A good sound man, a patriarch of the town,
the violent lights and slaves do not depress him,
he stands at ease on a far darker chasm
than that lost Carthage Scipio will burn,

lighting a cigarette in casual air.
Aeneas doesn't smoke: he has no vices.
The outer face reveals no inner crisis.
The children play around him without fear

who saw her once in a dim and crowded place.
His cry evoked indifferent hauteur.
Hell is as black as she is. That's her power.
And his to seek an emptier, lighter, place.

6

HAMLET

Sick of the place, he turned him towards night.
The mirrors flashed distorted images
of himself in court dress, with big bulbous eyes,
and curtains swaying in a greenish light.

Save me from these, he cried, I could not kill.
I did not have the true and pure belief
even to marry, reproduce myself
in finite mirrors, tall and visible.

Bad jokes and speeches, I endured them all,
so therefore let my death be a bad joke.
I see in the warped mirrors rapiers shake
their subtle poisons perfuming the hall

reflecting accidents, a circus merely,
a place of mirrors, an absurd conclusion.
Images bounce madly against reason
as, in a spoon, wide pictures, fat and jolly.

I could not kill, but let them have their deaths
imprisoned in this air in which they perish
where only lies and ponderous jokes can flourish.

Remove the mirror, for there is no breath.

7

EPITAPH

This is the Law. What you love best you get.
Position on the Council if you wish it.
Enough money never to be in debt
and to attract if not repay a visit.
Because your honour is most firmly set
in every action no one will dismiss it,
and lastly, when you die, they'll not forget
to buy a hasty wreath for your deposit.

If you had wanted greatness you'd have got it.
At least I suppose you would for some still do,
(especially the ones who most have sought it).
And when one thinks of you, so gross, untrue,
and then of failures from real greatness parted,
why then they wished it less than smallness you.

8

BURNS

"Ah such a genius," the church intones.
"Such love of justice and equality.
An ordinary person just like us.
Flesh of our flesh, bones of our very bones.
A genius walking under an Ayrshire sky.

Such pity and such melody in his songs.
A horny-handed son of toil and pasture.
Even in Russia they know of him,
even that country of such manifold wrongs,
immune to charity and even to culture,

worships our simple Burns, our dearest Rabbie.
He may have sinned. Which of us hasn't sinned?
But which of us has left such gems behind him?
So therefore let us toast him, wife and hubby,
giving thanks to God for the treasure of his mind."

And somewhere in a windy cornfield lying
under a heaven varying with its clouds
you touched her breasts to brilliant inspiration.
Her naked legs spun stanzas and her crying
was more insistent than was ever God's

in a far colder Eden than you now
enter in scent of hay and autumn grasses.
Her closing eyes adore you as you lie.
Son of the plough, you knew how best to plough
no barren land endeared by loud jackasses.

9

LENIN

You heard his voice, that terrible blunt voice.
His train was smoking in the station still.
He was so bald, so competent, so small.
Behind the words a beast was crying "Rejoice".

Snow palaces were melting as you stared.
The roads took one direction. And his head
clanged like a bell. "Let the dead bury their dead."
In his vast temple one vein blued and stirred,

and then became a sword piercing your heart.
You cried "Rejoice. I am the victim now."
"Blood," he was shouting, "blood. To grease the plough."
For like yourselves he did not care for art.

'Das Kapital' turned bronze within his fist.
There were some gouts of steam. That train had poured
a dreadful plague across each town and yard.
It was destructive and yet wholly blest.

"I am the victim. Why should I rejoice?
I've waited for my killer all these years,
for the assassin in his streaming hearse
to rape my garden and destroy my house.

He has the courage to uproot the chairs,
expel the sofas, pull the carpets up,
rouse my drugged wife from her suburban sleep
and send his lightning through my somnolent airs.

My hands applaud my death. I see the wolf
lope through the steppes with eyes of brilliant green
In this new spring I'm victim in the scene.
My pale fists tighten as I play at golf."

10

All night the poet drinks and all the day.
Disorder is the order he adores.
There are such quarrels and such slamming doors.
Such broken saucers and such debts to pay.

Sleep after drink and then the feeding eyes
circling his wasted moon, demanding light.
An eternity of talk and chains of wit.
After such foolishness the heart is wise.

Voices and taxis and the swaying stars,
the serious graduates expand their theses.
"Oh all I am is a hypothesis.
I haunt like Tennyson the harbour bars."

There is a man who's tethered to his dog,
and cleans his teeth in a round morning mirror,
who weeps all day for the minutest error
and snores from tidy pillows like a frog.

To other men the ambulance is speeding,
to other homes fire engines clang with bells,
for other men great whores inspect in halls
their planetary faces without pardon.

The poet longs at last to enter this
world without waste, this tidy hospital,
where all his images at last are still
and every error has a perfect price

where the smooth cat is roaring like a lion,
the wife is mistress to a thousand kings,
and from the patterned ceiling howl and sing
mowers transformed to tanks, and pens to talons.

11

It was the heavy jokes, the dreadful jokes,
the pewter-coloured jokes that drove you mad.

It was these jokes, the ornaments of zero,
the squat sad china made you die of shame.

It was the jokes that know no airiness,
immune to seasons, drove you to your knees.

It was the spoon-faced jokes, the Hall of Mirrors,
the circus air that made your speech opaque.

It was the flat-faced men with silver borders
of gold and silver made you live in fire.

It was their speeches robbed you of your speech
and caused pure silence to reflect your love.

The guns with crooked sights brought no reward
but pewter plates and a dishonoured word,

a doll shaped like Ophelia and a vase
vulgar as kings, as Claudius obese.

12

My Scottish towns with Town Halls and with courts,
with tidy flowering squares and small squat towers,
with steady traffic, the clock's cruising hours,
the ruined castles and the empty forts,

you are so still one could believe you dead.
Policemen stroll beneath the leaves and sun.
Pale bank clerks sit on benches after noon
totting the tulips, entering clouds in red.

And yet from such quiet places furies start.
Gauleiters pace by curtained windows, grass
absorbs the blood of mild philosophers.
Artists are killed for an inferior art.

Mad bank clerks bubble with a strange new world.
Insulted waiters take a fierce revenge.
Inside expanding cages aesthetes lunge
at cripples or despisers of the word.

Stout fleshy matrons send their pekinese
on wolfish expeditions and the night
is palpitant with howls. The scraggy throat
of some schoolmistress sways in a new breeze.

The butcher's hairy hand raises an axe.
White heads are neatly sliced like morning bread.
The errand boy rides whistling through the dead.
The scholar hacks at documents and books.

And skies are clearer than they ever were.
The haze has lifted. Fiction becomes fact.
Desire has seized at last the virgin Act.
And distant Belsen smokes in the calm air.

13

When you were young, my dear, you could be found
at barricades bristling in the morning mist.
Who could you not quote then? You were the host
to all the guests that agitate the mind,

dressed in ironic cravats, buttonholed
with roses of a future revolution,
the geniuses of terror and seclusion,
users of lead and quite immune to gold.

But now you say, "Such nonsense it all was!
My silly youth! One must mature, you know!"
And so you have matured. And so. And so.
For after lead and gold there comes the brass.

Ah, it is difficult. I know it well.
But surely it is possible to remain
a spy within the country and to gain
a hard-won honesty from hollow hell.

At least not look as if you can believe
that such a world is true, that speeches can
create not just an image but a man.
At least have the compassion still to grieve

if only for yourself and what you were
a truer student than a graduate
who wear your tie at your own dwarfish gate,
asthmatic traitor to a fresher air.

14

In youth to have mocked the pompous and shot down
these busy cruisers from an infinite sky
was such a sport as good as wittily
to shoot bon-mots at films. It seemed the moon
and the raw air justified utterly

all brilliance, all scorn of fat proud men,
pursing their names from newspapers, on stages
trapped by the fierceness of our tingling rages,
for after all they could not know of pain
but lived on vanity and subterfuges.

To think that, pelted by our quick depth charges,
zigzagging and crisscrossing in that sea,
they should survive, should cruise eternally
was not within our thought, for all the ages
had taught us surely that the boring die

and that the salon is the lasting room.
But that it isn't so our mirrors tell,
preparing early an inscrutable
face to outface the daytime and our shame,
and thinking too we'd have been capable

of the wittiest deeds if these had not before us
walked with the same grave pace as we do now
in self-protection from the gangs who prowl
mindlessly our actions in slow chorus,
foes of those burdened by "the true and real".

FAREWELL PARTY

Oh, it was such a party. Miss MacMillan
was sitting on the floor in stockinged feet.
The sherry brimmed and as for the red port . . . !
Wouldn't have missed it for a hundred million.

We sent her off in proper style I tell you.
Our Mr Morris (just imagine that)
sang a few verses, wore a paper hat.
It was, you won't believe it, a bright yellow.

Oh, such a time we had. And Mr Reilly
told such good jokes. I never laughed so much.
I really ached. And you should see him lurch,
drunk, he said, in the Army at Reveille.

Of course he was in the War. He can't forget.
Anyway it was midnight when we parted.
I felt so lonely and so broken-hearted.
I wish the party had been going yet.

But anyway I reached my room at last.
I'm not ashamed to tell you I was crying.
She served for fifty years, you know, Miss Lyon.
And all she said was, "How it all soon passed."

But I was punctual at my work next day
though a bit headachy, you can imagine.
But it's a memory for an ageing engine
as it steams on. That's what I always say.

16

CHURCH

Once more you speak, once more I hear your words.
The lights are bottle-shaped and milky white.
A blue flared angel spreads her wings straight out
from common oak matured in common woods.

Dressed in its Sunday best the club's assured
that God presides at their communion still,
holy and perfect and invisible
and punctual as the watch that lies obscured

in an ironed waistcoat on its golden chain
ticking so comfortably, taking no queer leaps,
but steady, measured, in its clear eclipse,
golden and round and perfect in design.

The drunk may howl by lost and broken quays,
the heathen fall in battles while our guns
stamped with His name exact the perfect silence
which laps this building in perpetual peace.

Among the tombs they shake pale hands and smile.
Laughter's too loud. Tomorrow, holy trade
(blessed by the Lord) may find these two arrayed
at opposite desks like navies poised to kill.

But now the leaves wink peacefully, and green
these mottled faces with the season's growth.
Below them bones, once furious with breath,
lie in predestined patterns. God must mean

success and harmony, his dividends
are still increasing and to get your share
all that you need is faith. His interests are
world-wide and deep. And these remain his friends

who will adore his every motion and
flatter his angers, call him competent,
worship his vast possessions and consent
to be his servant, pliant to command,

suave and collected when his thunders speak
and the glass door is shaken by his rage
so that his name is swaying, till the purge
of useless workmen will appease his pique.

The storm's invisible but rational.
The birds are singing in the graveyard now.
The world is perfect and on every bough
the golden angels shine and reconcile.

17

The wind roars. Thousands of miles it came
from Biafra or Vietnam. I lift my spoon
and see in it faces without hope or name.

I lift my spoon and lay it down again.
The mouth works forward. The high wind's a drum
lashing my body into fog and rain.

It is a dream, I cry, it is a dream.
The mouth breaks on the spoon with frenzied gum.

18

The zebra's running free. Then the wild dogs,
swerving inside, are clinging to its flanks,
slowing it down, devouring, in grey links
of long sharp gnawing teeth. In massive shock

the zebra stands, dead-stopped, milked by the fangs
which draw not milk but blood, great hunks of flesh.
One dog is leaping madly at its nose.
Its round eyes are locked in staring rings.

The plain is silent but for sucking yelpings
and the safe herd withdraws to the horizon
where the great suns of God intensely burn
above the long slim shadows blackly galloping.

19

YOUNG GIRL SINGING PSALM

Just for a moment then as you raised your book—
it must have been the way your glasses looked

above the round red cheeks—as you poured out
the psalm's grim music from a pulsing throat,

that moment, as I say, I saw you stand
thirty years hence, the hymn book in your hand,

a fleshy matron who are now sixteen.
The skin is coarser, you are less serene.

What now is fervour is pure habit then.
To bridge devoted and to thought immune,

a connoisseur of flowers and sales of work,
you cycle through round noons where no sins lurk,

your large pink hat a garden round your head,
the cosy wheel of comfort and of God.

And, as I see you, matron of that day,
I wonder, girl, which is the better way—

in innocent fervour tackling antique verse,
or pink Persephone, innocently coarse.

20

NATIONAL SERVICE

A small neat man, bright as a silver buckle,
he strode across the barrack floor, then stood
straight as a quivering dagger in the wood.
Then he began to speak and his hard will
held our pale eyes. He said: "You look no good,

a no-good bunch I've got to make soldiers of.
Ten weeks we've got, no fricking more nor less.
When I say 'Move' you'll move your fricking flesh.
You'll hate my guts. You'll say 'I've had enough.'
But you'll get more. That I sincerely promise.

I want no mother-suckers in this squad.
I want the best. You're out of bed at six,
you're on parade at eight and no smart tricks
like writing your MP, and saying 'It's bad.'
Cos if you do you'll feel a ton of bricks

descending, university friends. No questions?
You can ask me any question within reason.
You're in the Army now and I'm the person
to make sure you know it, so you'd better listen.
And no loose nattering. Here we call it treason."

How beautiful the day was. Past the square
the random birds flew easily in the sky.
"Man must conform," I thought. But then: "So they
also conform, by instinct". What we are
is the dreadful price the human mind must pay

for its own freedom. And I heard him turn—
boots clattering on wood. I laid my case
on the regulation blanket. There were plays,
poems and novels, and I pushed them down
to the very bottom, deep below the clothes.

There was no time. When ever is there time?
If what I am will fail, then this fails too.
What we have read may teach us what to do.
In this bare place they teach us how to rhyme
but not to state. The statement is what's true,

what we can name between the rhymes, what judge.
Not to loose harmony to give our heart,
or prompt the innocent with words that hurt.
And these great absences which are our badge
are also the great absences of art.

21

AT THE SALE

Old beds, old chairs, old mattresses, old books,
old pictures of coiffed women, hatted men,
ministers with clamped lips and flowing beards,
a Duke in his Highland den,
and, scattered among these, old copper fire-guards,
stone water-bottles, stoves and shepherds' crooks.

How much goes out of fashion and how soon!
The double-columned leather-covered tomes
recall those praying Covenanters still
adamant against Rome's
adamant empire. Every article
is soaked in time and dust and sweat and rust. What tune

warbled from that phonograph? Who played
that gap-toothed dumb piano? Who once moved
with that white chamber pot through an ancient room?
And who was it that loved
to see her own reflection in the gloom
of that webbed mirror? And who was it that prayed

holding that Bible in her fading hands?
The auctioneer's quick eyes swoop on a glance,
a half-seen movement. In the inner ring
a boy in serious stance
holds up a fan, a piece of curtaining,
an hour-glass with its trickle of old sand.

We walk around and find an old machine.
On one side pump, on another turn a wheel.
But nothing happens. What's this object for?
Imagine how we will
endlessly pump and turn for forty years
and then receive a pension, smart and clean,

climbing a dais to such loud applause
as shakes the hall for toiling without fail
at this strange nameless gadget, pumping, turning,
each day oiling the wheel
with zeal and eagerness and freshness burning
in a happy country of anonymous laws,

while the ghostly hands are clapping and the chairs
grow older as we look, the pictures fade,
the stone is changed to rubber, and the wheel
elaborates its rayed
brilliance and complexity and we feel
the spade become a scoop, cropping the grass,

and the flesh itself becomes unnecessary.
O hold me, love, in this appalling place.
Let your hand stay me by this mattress here
and this tall ruined glass,
by this dismembered radio, this queer
machine that waits and has no history.

22

Could I but love you then your hands are green,
the wolves and mice come streaming out of heaven,
and all the land from Islay to the Leven
becomes at that strange moment less serene.

In peaceful pools the deadly pike are striking,
the deer lock horns on the tree-locked mountain side,
the weasel bites into the rabbit's head
and the great stones in mountain streams are breaking.

O Death how brilliant is your round green eye
shining above Ben Cruachan. In old valleys
the fox is loping with his grinning malice
and your warm eyes are merrier than the sky.

23

Take, O take that book away.
We know by instinct a far other grammar.
On a warm night of a true Scottish glamour
we stood together by a Scottish tree.

The moon was mild, the sheep peacefully gnawing
the late wet grass, as bright as hunting eyes.
The tree was old. I knew it by its size.
Ring on vast ring it had been long a-growing.

All love is hungry and your silken dress
thin as an insect's wings when it is flying,
its mate beside it. You can't hear its crying,
if cry it does, expending its brief fires.

The tree was old with bark. Dear love, consider
how from this awesome circle we will pass
into indifference and bitterness
though, ring on ring, your hair enchants the water.

24

More than twenty years ago you heard it
in a far desert tantalisingly floated

towards your tanks and tents that drifted white
in that voluminous gritty skin of light—

"Lili Marlene" in German. By a gate
standing in whorish raincoat, cigarette

dangling eternally from the schoolgirl lips
she waits in lamplight while the guard half sleeps

leaning on his rifle. Now she snores
beside you in your terrace house. There roars

a car accelerating and on the ceiling
its lights converge and cross, a searchlight feeling

for some invisible enemy. In its cot
your swathed child sleeps, so vulnerably white,

the moonlight on its skull, a small thin dome.
A helmeted Junker thunders round your home.

The umbrella collapses on pearl-headed sticks
and blooms at morning to a tent. What tricks

that belly dancer had! How cool mess tins
in these barbaric Mediterranean dawns!

And, late at night, that taunting German voice.
Oh, how you miss it, tense with tenderness

among the bowler hats, the frozen 'Times',
Lili Marlene the schoolgirl of sweet rhymes

standing forever in her raincoat near
the fading lamps of martial leather gear

now rotting as you watch another dawn
reveal your wife camped in her flesh and bone

and, faintly past the lawn so flat and green,
a mocking voice that sings 'Auf Wiedersehen'.

25

I take it from you—small token of esteem—
this ponderous watch that holds a soundless scream

and give you back—a gift that haunts your sky—
the howling faces of eternity.

26

I don't love you. You have surrendered too much,
love, conversation, everything but your rich

honour and dignity—hair that's powdered white
shines, as you think, like high stars in the night

and the red ring on your finger signifies love
(so say the Supplements from their ageless grove)

Ah, witchlike oracle with the stringy neck
far from the nereids, conscienceless and Greek!

And I, I say you've surrendered far too much
who am a scream only, from the porch,

the Stoa, looking out as Poor Truth passes,
and then yourself in beady demon glasses.

27

Shyly she said: "I so admire your work.
It's as if you really knew what I was feeling.
All art is universal, don't you find?
And wasn't it the poet William Blake
who wrote that bit about the grain of sand?

Would you agree the poet has a talent
for empathy—at least I think it's so.
I feel it with Eliot, Auden not so much.
Don't you distrust the falsely brilliant?
This fashionable craze 'to be in touch'?

I'd love to hear your views"—
                       And what I see,
blindingly accurate, is a great bare church
and, in it, spiritual artists middle-aged
or—if they're young—intently visionary,
gap-toothed and glassy, ugly and unrouged.

And where, I ask, are all the beautiful?
They're never seen at seance or at church.
Nor yet at poetry readings are they present.
In the underwater light of Sunday School
only the pale and moderate are lessoned

and then go off to war with faint bowed heads.
In barbarous sunlight, crops swinging from hands,
the helmet-headed dazzling riders go
while clerkish conscripts and their holy brides
chatter their Yesses from a deeper No.

28

To hell with this poetry reading, he cried—
your faces that are clocks within my side—

It's you my works supposed to be attacking
so why in God's name don't you cease your clapping?

29

The hall is large and echoing. You are small.
Speech after speech. I will not tell you all

my heart is thinking of. You say, "The rates
are far too high." And then, "Suggested sites

for the new hall are really quite absurd."
Speech after speech. The nightingales once heard

Socratic dialogues but did not know
the place was a loved prison. And the slow

poison beat its oars out on Death's stream.
I might have asked a question if the dream

had not so held me of the rooted state
the weight of leaves, the tentacles of light,

that where we love we stay in spite of all
and why we love, no burrowing mind can tell.

30

I know you, Hawthorne, I know all your ways.
New England isn't different from this

place I might call Old Scotland. There's the stiff,
formal, bristling, fence round brimming life,

the cultural silence and the open sea
where the assassins cruise though it seems free.

Your books are dossiers only, flawed by what
you seek to spy on, the most brilliant art

of an unnatural garden, where your Eve
perishes from placards. I believe

in other writers (Dickens is the type)
who live in unjudged profusion, half asleep,

wandering among people with real teeth,
real rotting livers, comedy and death,

Pickwicks not broken by intelligence
but flourishing like poppies and as dense

as any real plant is, systematic
only in craziness and immune to critic

or spy with shiny briefcase. It is he
who loves the untended garden and the sea

with its manifold varieties of killer,
green wave, green plant, green envy and green squalor,

whose fogs though evil are yet real fogs
whose loves emerge from real London bricks

whose crazy worlds collide as dodgems do

and burst like fireworks in the comic blue.

31

The lawyer stands beside the sea
humming to himself. Eternity

is a large white document before him.
In His Will is his decorum.

The sea is rather extravagant and defiant
but nevertheless God is a cunning client

with an eye for the small print, a love of fences,
brutal litigation, tough expenses,

and, above all, a language more obscure
than even a lawyer's.
                              The great waters pour

at his small dapper feet. He feels at home
by the sea's phased formalities. A dream

of happy carbons repeats busy Nature
on an immense harmonious typewriter.

32

RETIRAL

1

You say goodbye and hand the banknotes over
in the flash bulb light that heavens you forever

and then he goes into the world Of Time,
a moderate man unschooled by the extreme.

2

Dabbling in gardens, making lectures up
for the Local Ladies Literary Group,

he fights the staring second and at bowls
assembles nightly with some damnèd souls.

3

We meet him at vague corners. He converses
on nineteenth century education courses,

a moral man dismissed to ghostly earth.
The wall of Time has pictures of his death.

4

Let no one speak of this but still pretend
it's a beginning when it's just an end

as skeletons make music in the wind
and crack these dreadful jokes that break the mind.

5

Doomed to two lines of print he mows the grass
whose recitation was "Amo, amas."

The jukeboxes flash out no bright M.A.
though purple as his gown their colours sway.

6

God stole his children from the deep-carved desk.
Man praised his conscience, his accomplished task.

On tumblers and on aspirins shine the stars.
He cannot count them and he cannot parse.

7

Analysis has failed to rule the globe
with its tremendous and luxuriant hubbub

but God's vast register will find him still
the most obedient pupil in His school.

33

What were you in search of, RLS,
peace for your rotting lungs or something else?

Genius of boys' books, your natural style
flowered best in them, your heroes found in Mull

spontaneous habitat, and the South Seas
harboured in turn purer simplicities

(waves that bounced like the dishonoured cheques
of unshaven men among the slummy docks).

White-helmeted sahib of a simpler race
what were you looking for? Oh, not for us

these pretty oceans with their happy smiles.
Your boyish books belong to older Mulls,

your style broke on the adult where we start.

Think of the rope trick, sahib, that's our art.

34

What's your Success to me who read the great dead,
whose marble faces, consistent overhead,

outstare my verse? What are your chains to me,
your baubles and your rings? That scrutiny

turns on me always. Over terraced houses
these satellites rotate and in deep spaces

the hammered poetry of Dante turns
light as a wristwatch, bright as a thousand suns.

35

It is the night, night of the yellow moon,
night of the yellow lamplight, yellow stone.
The yellow dog runs home with his yellow bone.

We are infected. We begin in red.
We shake our tiny fists against the dead
and cry new words, and clamour to be fed.

On the yellow street the yellow drunk goes by.
His yellow face receives the yellow sky.
Not the red light will shake him utterly,

Nor of the large young cars the memory.

36

Love, if you had been love you would have spoken.
I am so weak. Tell me the world is good.
Offer me your goodies. Misbehave.

I am so frightened of my visions. Hold me.
Let me rest in the ruins of your arms.
Let your eyes flash a clear unbourgeois honour.

Tell me that for some cause the soldiers died.
That they who went out rippling (came home gulping)
breathe an oxygen that ennobles them.

Star of the low west, our liberal light,
I'm frightened of your going. I'm as weak
as the clouds that straggle round its fallen flag.

Enmesh me, love, in love. I hate to hear
the creaking tenement, the blind man to see
whose legs are stumps, and yours so white and round.

37

Your face is a white moon rising
over Islay,
over Mull, Tiree of the corn.
The branches of Latin it has outsoared,
the complications of the conjugate verb.
The white moon rises clear of everything.
It has achieved what cleverness has not
(under the earth there run the secret streams).
Let not the green ghost by the tombstone,
the bibles, open in their perfect granite,
let these not hold you down.
The moon rises,
achieved and white. Let not the heart be troubled,
believe in love, believe in perfect vases,
believe in gravity that moves in light.
I, among branches, looking up, am shaken.
On what is far beyond me my teeth close.

38

LITTLE BOY

Little boy, carefully stringing your bow
in a little village on a somnolent day,

playing alone, standing beside the road
in your copper-coloured shorts, your bow is red,

the arrow itself green. You fire. It goes
perhaps six feet. And you remind me so

of someone else—who is it?—that I test
some words to say to you: "In this sleepy waste

the arrows don't go far, it is the air,
the tall banked ferns that climb in stair on stair,

the people in their mansion-polished rooms,
the razored lawns, the roses' clammy fumes.

Go closer to the sea," I almost cry,
"Where the salty mussels are and the huge sky

reflects itself in water, and the sun
is orange at the limits. Fire it then,

your bright green arrow, Cupid of the waves,
across the rocks and sands and endless graves

straight at the sun, your target, and forget
this sleepy village, sunk in its ferns and light."

—I almost spoke remembering a lad
reared among the voices of the dead
a bow in his hand and music in his head.

39

Children, follow the dwarfs and the giants and the wolves,
into the Wood of Unknowing, into the leaves

where the terrible granny perches and sings to herself
past the tumultuous seasons high on her shelf.

Do not go with the Man with the Smiling Face,
nor yet with the Lady with the Flowery Dress.

Avoid the Crystal, run where the waters go
and follow them past the Icebergs and the Snow.

Avoid the Man with the Book, the Speech Machine,
and the Rinsoed Boy who is forever clean.

Keep clear of the Scholar and the domestic Dog
and, rather than Sunny Smoothness, choose the Fog.

Follow your love, the butterfly, where it spins
over the wall, the hedge, the road, the fence,

and love the Disordered Man who sings like a river
whose form is Love, whose country is Forever.